This book is for my mother,
who never met an animal she wouldn't take in,
or a plant she could not keep alive.

Featured Tiny Thinkers
Claire Stanton
Aiden Van Ostrand

Rachel Hears a Song
A Tiny Thinkers Book

Written by: M.J. Mouton
Illustrated by: Jezreel S. Cuevas
Edited by: David Smalley & Amanda Franquet
Book Design: Nathalie Wittmann

Printed in China

US Library of Congress
ISBN: 978-0-692-54827-1

Hi, I'm Hitch!

I've spent time with some amazing Tiny Thinkers! Join me as we learn about people and the science they discovered. And see if you can spot me along the way, as I tell you the story of Rachel's real-life adventure that changed the world!

RACHEL
HEARS A SONG

Written by M.J. Mouton   Illustrated by Jezreel S. Cuevas

A Foreword by
Dr. Mark J. Plotkin

Hi—

I'm Mark. I study medicinal plants in the rainforest.

But before I began learning about healing plants and before I began visiting the Amazon rainforest, I read books about the world around me and the people who protect it.

One of the first heroes who inspired me was Rachel Carson, whom you will learn about in "Rachel Hears A Song."

As a little girl, Rachel began studying animals and plants on her family farm in the state of Pennsylvania. After that, she started reading about and then studying the ocean and all the creatures of that watery world. She then began writing books about the importance of protecting nature, including a very famous book called "Silent Spring," that convinced millions of people to work together to conserve the Earth.

"Rachel Hears A Song" will show you how one little girl began a career devoted to helping nature and people, too. We need more people like Rachel, both girls and boys. Please read and enjoy this wonderful book and think about how you — like Rachel — can help make the world a better place for plants, for animals, and for all of us!

Mark Plotkin
Amazon Conservation Team
www.amazonteam.org

Rachel loved listening to all the birds sing.
The birds filled the trees and welcomed the spring.

"So many beautiful birds around our home,"
Rachel thought. "How boring would spring be if they
were all gone?"

The wrens' tweets were a thing of beauty.
The doves' songs were calm and moody.

She could listen to the robins' song forever,
but it wasn't perfect until they all sang together.

Even a crow's raspy caws went right along,
with the rest of the birds creating spring's song.

Rachel enjoyed the song on a day that
could not be much better,
as an airplane flew over, lower than ever.

"Why is this plane flying so low?" Rachel wondered. "Was it spraying the fields to help the farmer's plants grow?"

It never occurred to her that something was wrong, as the plane passed again, disturbing the song.

The next spring came and something caught Rachel's attention. "The birds' songs are not as loud," Rachel thought, as she listened.

The spring song of the birds had lost some of its beauty.

Either some were not singing, or there weren't that many.

Rachel heard a raspy crow trying to sing
his part of the song.

She approached him quite cautiously to ask him,
"What's wrong?"

Crow gave Rachel a very sad look.
"I think it's the stuff from the plane.
It's in the fields and the brook."

Crow said, "I know the farmers mean well, they are just taking care of their plants.

Their fields are being invaded by locusts and ants."

Rachel said, "Have they noticed
the birds are almost gone?
I will tell them myself!

Maybe they don't know
it does more harm than it helps.

Maybe there is something
that can work just as well.
We must save the birds!
There must be someone to tell!"

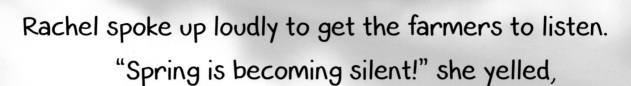

Rachel spoke up loudly to get the farmers to listen.

"Spring is becoming silent!" she yelled,
to get their attention.

The Crow flew in to help and offer his reason,
as to why there are so few birds singing this season.

"We have to protect the plants, animals, and the water. I think we can do it if we did it much smarter.

You can test what works and does not harm the Earth, save the animals and plants with something that works."

On the first day of spring,
Rachel was still in her bed.
Crow flew into her window
and perched by her head.

He said, "Winter is over,
the flowers are blooming,
the rabbits are out early
because something is brewing."

"All right," Rachel said.
"I'll see what is all the fuss."

Rachel stepped out of her house,
and the woods were totally quiet.

She started to cry,
as the spring that she loved
had truly gone silent.

Up in a tree the crow sat with a very sly look.
He released a note so beautiful and long
that the whole tree shook.

Rachel said, "Nice try Mr. Crow, but it is just not the same. I am so sad right now that I can't find the words.

To play the perfect spring song requires all the birds."

Rachel perked up as she heard a faint tweeting sound.

She looked at the tree, but no birds could be found.

Except for the crow who let out a note louder than before.

He winked at Rachel and asked, "Are you ready for more?"

The tree came alive with the flutter of wings,
as Rachel's eyes brightened up with the song of spring.

Every single bird from the
forest was there.

Every note that they sang filled
every space in the air.

From small birds, to large birds,
to the top of their ranks.

They sang this song for Rachel to
show her their thanks.

Rachel had done something that was done by no other.

She said, "Let's save the Earth, this miraculous wonder."

Rachel Carson did not want spring to go silent.

Her effort pioneered a method called environmental science.

It all started with one question.
Now we remember her work.

She asked, "We know how things help us,
but how do they hurt?"

Pay attention to the birds. We need them as much as they needed Rachel. Maybe you will be their next scientific hero!

Rachel grew up
to be known as...

Rachel Carson
1907-1954

Rachel Carson was born in Pennsylvania and raised in Springdale on a farm with her dog, Candy. Her mother, a school teacher, encouraged Rachel's love for reading, and nurtured her love for nature. Rachel started writing stories about animals at the age of 8 and published her first article at age 10.

Rachel went to college and studied biology and creative writing, along with zoology and genetics. After college, she wrote three popular books about oceans while she worked at the U.S. Wildlife and Fisheries.

Farmers across the country were using pesticides to rid their fields of insects. The pesticide DDT proved very effective against mosquitoes and other bugs. Rachel spent 4 years gathering examples of the damage that the use of DDT was causing. Birds including bald eagles, peregrine falcons, pelicans, and many song birds were quickly endangered.

Rachel's book Silent Spring brought attention to the environmental damage DDT was having. Rachel did not ask for DDT to be banned, but argued for the study on the harms that some chemicals could have on the environment. Rachel's effort and book Silent Spring are widely regarded to be the foundation on which the United States' Environmental Protection Agency (EPA) was founded. Her concerns and questions started an environmental movement.

On this day, those birds that Rachel worried about have made a significant comeback, but it is not over. Maybe you will be the next great scientist to carry on Rachel's work.